FIRE

destructive • regenerative • sacred

Metchosin ArtPod Gallery

Edited by Edward Peck and Phyllis Schwartz
Editorial assistance from Diana Smith
Copy editing by Cindy McMath
Designed by Phyllis Schwartz
Published by Sassamatt Publications
Front and back covers by Cindy McMath, Fire: *The Beauty and the Chaos*

sassamatt.com
ISBN: 978-1-06993244-1-4
SASSAMATT
publication

FIRE: DESTRUCTIVE, REGENERATIVE, SACRED

Fire is a force that orders and dissolves, comforts and menaces, cleanses and scars. Elemental in the truest sense, it is embedded in geology and weather, woven into ritual and story, central to both survival and spectacle. Artists in this exhibition approach fire from many entry points: as an agent of ecological trauma and recovery in the wake of wildfires; as a source of warmth, memory, and community at hearth and campfire; as ritual material and symbol in cultural practice; and as a metaphor for psychological, political, and emotional upheaval. Across painting, sculpture, installation, ceramics, photography, video, digital media, performance, and mixed media, the works gathered here make visible fire's power to transform landscapes, objects, and lives.

Curated by Khione Gardner, Kim Money, and Diana Smith, with guest juror Samantha Dickie, Fire: Destructive, Regenerative, Sacred brings together diverse aesthetic approaches and personal narratives. Some artists document aftermath and recovery, tracing charred topographies and the slow return of life. Others probe the aesthetics of flame and ash through abstraction and material experimentation. These works question, mourn, celebrate, and provoke. They speak about stewardship and the ceremonial roles of fire, reminding us that relationships with flame are culturally specific, ancestral, and ongoing. Other contributions turn inward, exploring grief, passion, and yearning, with fire as an intimate emblem of change and loss.

More than a record, this exhibition book is a space for reflection and conversation. The essay by the juror presents the ethical and ecological questions that fire raises, while artist statements bring us closer to individual creators. Together, these words and images shape memory, territory, identity, and the future of our shared ecosystems.

Phyllis Schwartz
Editor

Salted
Samantha Dickie

8 x 6 feet approx

Porcelain, soda fired, 2022

This constellation of ceramic works is a grouping that was initially part of a large installation of 1800 components that filled an enclosed room in a public gallery exhibition in 2022. This wall assemblage is a new iteration of this series. The variable colours and movement on the surface of these works are the result of salt being blown into a firing atmosphere where it is carried through the flame as it dances around the deep chamber of a high-fired soda kiln.

Samantha Dickie

Juror and Artist Statement

Fire, as the third element in this annual exhibition series celebrating Earth Day, is a multi-faceted, vital force. Each of the elements (fire, air, water and earth) are alive: pulsing, emerging, and sustaining life on the planet. They ask each of us to consider what it means to be human in a more-than-human world? How do we understand this element as a sentient force to be respected and to learn from? How do we think and act through each element?

Fire is comfort and danger, flow and explosion, renewal and destruction. Fire supports the balance of life, providing warmth and care, rejuvenation, and nourishment. The works juried into this exhibition are in conversation with each other, as they uniquely respond to the multiple characteristics of fire conceptually and aesthetically in both provocative and beautiful ways.

The works that I am sharing in this exhibition hold the marks of fire through the movement of smoke, flame and ash in various atmospheric kiln environments. As a ceramic artist, fire is not only integral to the artistic process, but it also becomes a gathering place for artistic sharing. Potters collaborate to fire an outdoor kiln, patiently tending to its needs with fastidious devotion through a slow and spacious process until the ideal temperature is reached for the desired surface. Firing a kiln is not just a community event but more a conduit for creative connection.

The conceptual arc of my practise is centred on attunement to the spaces in-between: an intentional quality of active noticing with curiosity, careful attentiveness, and slow wonderment as precursor to care and custodianship needed in the human and the more than human world. Attuning to fire is to watch it breath: drawing in, swelling out, dancing and resting, expanding spaciousness.

The poem, Fire, by Judy Brown beautifully illustrates this attunement.

What makes a fire burn
is space between the logs,
a breathing space.

Too much of a good thing,
too many logs
packed in too tight
can douse the flames
almost as surely
as a pail of water would.

So building fires
requires attention
to the spaces in between,
as much as to the wood.

When we are able to build
open spaces
in the same way
we have learned
to pile on the logs,
then we can come to see how
it is fuel, and the absence of fuel
together, that make fire possible.

We only need to lay a log
lightly from time to time.

A fire
grows
simply because the space is there,
with openings
in which the flame
that knows just how it wants to burn
can find its way.

Judy Brown, 2013

Stoked

Samantha Dickie

16 x 16 inches

Porcelain, wood fired, 2022-25

These small ceramic stone-like forms are the result of 60 hours being tumbled, poked, and prodded in the firebox of a wood kiln. Commonly described as pyromaniacs, a community of potters will gather for multiple days to hand-stoke chopped wood into a kiln, while attuning to the pulse of the fire to keep the temperature climbing up to over 2345 degrees fahrenheit. This is no small task. Stoked is the culmination of four years of woodfiring on Denman Island each spring. The fallen ash coats the pots as the flame courses through the large Anagama chamber, and only a few of these small forms fit into the firebox. Caked and crusted with ash, these precious ceramic stones hold a quality of geologic time; the imprint of fire from its deepest roar; the permanent mark of combustion.

Smoked.

Samantha Dickie

7 feet x 28 inches

Ceramic, saggar-fired, 2019

Each individual pod in this installation of slip-cast porcelain components holds the unique imprint of smoke as the carbon interacts with metal, salts and oxides in a low fired atmospheric barrel kiln environment. No two smoke fired pieces look alike.

Stacked
Samantha Dickie

3 feet tall (photo by Cathie Ferguson)

Ceramic, saggar-fired, 2019

These three foot tall, wheel-thrown bottles undergo a magical transformation inside a large gas-fired kiln. A custom built chamber is assembled around each of the sculptures to create a localized, individual, oxygen deprived environment at a controlled temperature. The fire, starving for oxygen, pulls it from the clay, creating specific colour changes in alignment with the various oxides and metals introduced inside the saggar.

Artist Biography

Based on the traditional territory of the Songhees and Esquimalt Nations in Victoria British Columbia, Samantha Dickie is a contemporary ceramic installation artist with two decades of experience executing a wide range of site-specific projects in public, corporate and private realms. Her architecturally scaled, multi-component installations have been exhibited in solo and group public exhibitions across Canada. Samantha holds an MFA, a BA in Gender and Indigenous Studies, as well as a Diploma of Craft and Design in Ceramics. She has attended artist residencies in France, China, Yukon, and Alberta. Samantha has been the recipient of national and provincial grants, local awards, has been the subject of numerous articles, reviews, and databases. Her work can be found permanently installed and in public collections across North America.

Website: www.samanthadickie.com

IG: @samantha.dickie.studio

VIRIDITAS
KEVIN KELIHER

18 x 22 x 1.5 inches

Digital collage archival ink print on 100% cotton rag

This digital collage draws inspiration from John Vaillant's 2023 book "Fire Weather", which discusses the 2016 Fort McMurray fire and pyro-tornadogenesis, as well as his depiction of Hildegard of Bingen—a twelfth-century Benedictine nun, artist, and naturalist. Hildegard's concept of "viriditas" (greenness) refers to the life force that enables renewal, symbolizing Earth's persistent ability to recover after disaster.

Container for the Flaming Words

Derek Kasper

24 x 20 x 20 inches

Stoneware clay and glazes

I am a practical person, and being appalled at the inflammatory language thrown about in conversation and by purported leaders of our world I thought I would create an octagonal ceramic receptacle that would be capable of containing all that fiery language under a locking lid. You may be able to notice some incendiary language trying to force itself out of containment, forming itself into words on the facets of the containment vessel.

Bumblebee Jamboree
Kim Money

20 x 20 x 1.5 inches

Acrylic, house paint, spray paint and paint pens on cradled panel

I was thinking about how peaceful and beautiful a huge bonfire can be. I wanted to portray a joyful atmosphere that was warm and inviting. A jamboree is a large festival gathering, full of music, bonfires and celebration. Part of the painting made me think of bees happily flitting around at their own jamboree and I added the honeycombs, so it became Bumblebee Jamboree.

Raining on Fire

Pam Maclean Belcher

12 x 24 x 1.5 inches

Acrylic on canvas

I first painted a post-fire landscape after a canoe trip in Northern Saskatchewan took us through a burnt-out part of the Churchill River system. The forest around us was heavily blackened, but the ground was carpeted with fireweed, in a vibrant pink and green contrast of renewal and resurgence. Since then, I have painted many forests fully on fire, and after the fires are out and the regrowth begins. Raining on Fire came during a really hard spring 'fire' season. Huge, all consuming fires were burning across many provinces, and then, to an almost collective sigh of relief, it began to rain. In a country as vast and forested and wild as Canada, we have to live with nature. So many of us have experienced what a wildfire means to the land around us. This painting is an acknowledgement of that cohabitation, and of the hope we continue to have that forests need to continually change, rain will come, and renewal will follow.

Vase

Laura Keil

6 x 4.5 x 4 inches

Wood-fired ceramics

This clay vase was wood fired in the Tozan Society's train kiln on Vancouver Island. Over a 30-hour period of continuous stoking, flames draw through the kiln leaving a natural ash glaze and marks showing their trail on the pots. It is a dance of the elements of fire and air that are recorded on the clay. Every piece is unique, and the result of a connection to the earth, the tree and the fire. The vase is placed in the ember pit of the kiln. As the hours of firing pass, the embers build up, surrounding the pots. The embers are the heart of the kiln - they emit a deep glowing, pulsating heat. The results of this process create an effect on the pot akin to an ancient object that has weathered many storms.

Lii Capote Flaamesh (The Spark Coat) (honourable mention)

Erin Macklem

65 x 24 x 18 inches

Mixed media using wool blanket and yarn

Creating this Métis *capote* from a Hudson's Bay blanket meant working inside a complicated history. The capote carries Métis stories of movement, survival, and ingenuity, while the blanket itself symbolizes both connection and deep harm tied to colonization and the systems that reshaped Indigenous lives. When I found a vintage blanket scarred by a campfire spark (burning the iconic label and lowering its monetary value) it felt aligned with the story I wanted to tell. The scorch mark echoes the damage done and the decline of that destructive empire. By cutting and reshaping the blanket I transformed this colonial symbol of imposed systems into a beautiful and practical garment that embodies Métis resilience and creativity. *Lii Capote Flaamesh* is a conversation between past and present—a remembering and a re-making. It loudly sings joyful campfire songs while whispering truths that sting like smoke in your eyes.

Ceasefire
Anna Mar

24 x 18 x 1 inches

Mixed media on canvas

A meditation on internal devastation. This piece explores the aspirations, futures, and potential of people caught in the flames of conflict. It captures the innocent lives that should have been vibrant, now lost to fire and fury.

Fire: a warning at a distance (Juror's Award)

John McFetrick

11 x 14 inches

Photograph on archival paper

Fire announces itself as prophecy. It arrives first as colour and heat, a distant signal softened by water and sky. Destructive, regenerative, sacred, it asks for reverence before resistance. Gold seduces. Blue thins. The warning reflects and delays, mistaken for beauty. This becomes a meditation on urgency. Distance breeds helplessness. The fire remains far off, yet everywhere implied. It waits. It always waits. By the time it takes the ground, the climate has already turned, and the future burns quietly in place.

Success at Last!

Derek Kasper

16 x 11 x 11 inches

Medium fire porcelain and glazes

I am amused by and interested in the historical interest in alchemy. By the same token I am interested in turning semi-random mixtures of toxic chemicals into the appearance of precious metals by simply applying a few thousand degrees of heat. Spinning a little bit of putty-like porcelain into a form deserving of the golden treatment is another sort of physical alchemy altogether.

Wildfire #1

Claire Kujundzic

9 x 18 x 0.2 inches

Mixed media on canvas

I began this series of paintings in the summer of 2021 while charred pine needles and ash fell from darkened skies over Wells over several weeks, and completed them this year. When we moved south to Victoria, we took a long detour to avoid Lytton which was on fire. Even from a far distance, we could see the tall plumes of smoke rising above the town. Using various media, including gold leaf and torn canvas "trees" attached to paintings allows me to depict aspects of my experience of witnessing large infernos at close hand. Lodgepole pines evolved with fire; their cones don't open without sufficient heat. Although some fires burned so hot that they vaporized all organic matter, many cones opened and seeded new life.

Raining On Fire

Pam Maclean Belcher

12 x 24 x 1.5 inches

Acrylic on canvas

I first painted a post-fire landscape after a canoe trip in Northern Saskatchewan took us through a burnt out part of the Churchill River system. The forest around us was heavily blackened, but the ground was carpeted with fireweed, in a vibrant pink and green contrast of renewal and resurgence. Since then, I have painted many forests fully on fire, and after the fires are out and the regrowth begins. Raining on Fire came during a really hard spring 'fire' season. Huge, all consuming fires were burning across many provinces, and then, to an almost collective sigh of relief, it began to rain. In a country as vast and forested and wild as Canada, we have to live with nature. So many of us have experienced what a wildfire means to the land around us. This painting is an acknowledgement of that cohabitation, and of the hope we continue to have that forests need to continually change, rain will come, and renewal will follow.

I'VE TASTED DESIRE
KEVIN KELIHER

14.5 x 17 x 1.5 inches

Watercolour on Arches 140 lbs cold pressed cotton rag

Some say the world will end in fire, Some say ice. From what I've tasted of desire I hold with those who favor fire." Robert Frost, "Fire and Ice". The consequences of burning millions of years of accumulated fossil energy in a period of decades will be ongoing & dramatic. (Vaillant, 2023)

Dreams on Fire
Anna Mar

24 x 18 x 1.5 inches

Mixed media on canvas

A meditation on internal devastation. This piece explores the aspirations, futures, and potential of people caught in the flames of conflict. It captures the innocent lives that should have been vibrant, now lost to fire and fury.

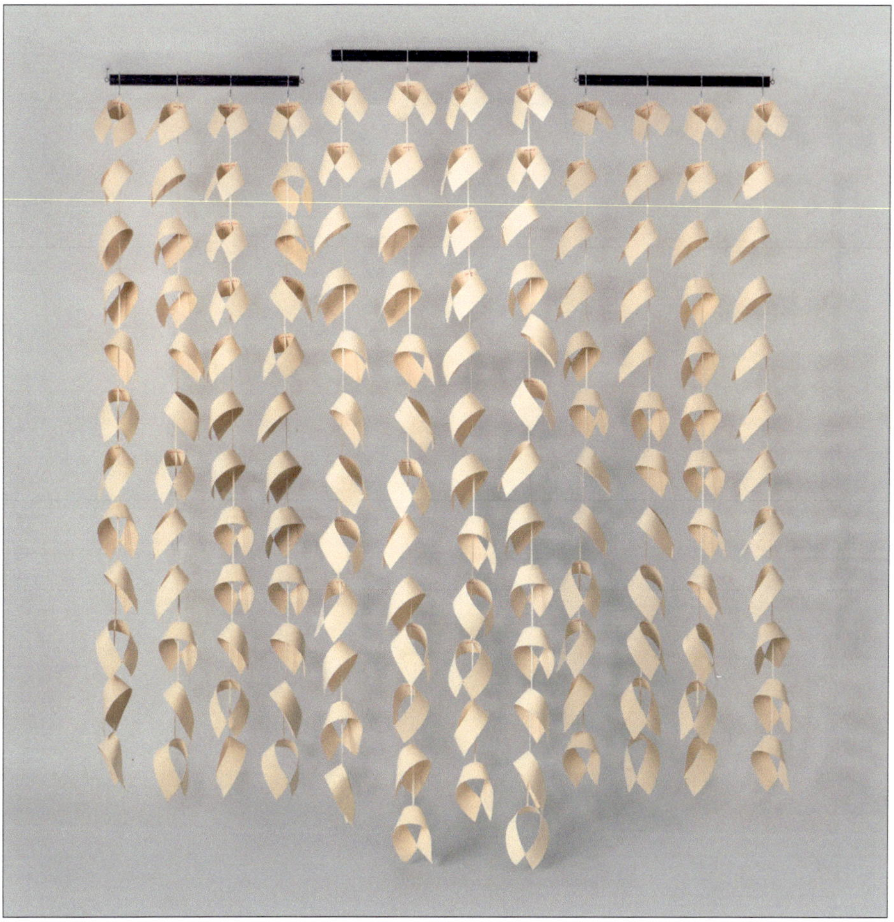

Triangle Shirtwaist Factory Fire 1911 (Juror's Award)

Barbara McCaffrey

85 x 72 x 12 inches

Cotton, cotton canvas, factory pattern hangers

On Saturday, March 25, 1911, 146 garment workers, of which 123 were women and 22 were men, perished in the Triangle Shirtwaist Factory Fire. Most of the workers were Italian or Jewish immigrants aged 14-25, were paid pittance and worked long hours. The fire was said to have started with the dropping of a cigarette into the remnants on the floor, which ignited in a flash. The owners, in their attempt to keep the workers at their stations, had not provided water buckets at the doors and even those were locked to prevent employees from stealing during the work day. This is a quiet piece. No flames, visible devastation or noise that would give you the impression of the scene. It is a memorial to the women and men who lost their lives seeking a better future, but who succumbed to the greed of the men who owned the building and business. Each of their names are printed on the inside of the collars.

Fire: the Beauty and the Chaos

Cindy McMath

30 x 24 x 1.5 inches

Acrylic on canvas

I've always been fascinated by fire. This painting features both the beauty of fire, and its chaotic, unpredictable nature. Fire fills us with wonder, at the same time it demands our respect.

Wood Wide Web
Kim Money

10 x 8 x 1.5 inches

Acrylic and collage

There is evidence that trees communicate through their roots using a network of fungi called the mycorrhizal network. They can warn each other of potential threats like drought, disease, insects and possibly fire. This communication system is called the Wood Wide Web and beyond warnings they share nutrients, water and carbon, supporting sick or young trees.

Leo Rising
Cindy McMath

24 x 8 x 1.5 inches

Mixed media on cradled wooden panel

This piece represents the fire sign, Leo. Named after the fact, once I saw the (very) abstract lion I couldn't unsee it.

Evoking Memories of Fire

Edward Peck

17 x 22 x 1.5 inches

Pigment ink on cotton paper A/P

There are fires in the summer, high seas in the winter that eat the coast, the seasons mix into each other, the garden produce is erratic, there are creatures from away… We see the chaos sloughing toward us, "everywhere the ceremony of innocence is drowning. The best lack conviction, while the worst are full of passionate intensity." Yet we go on "turning and turning in a widening gyre. The falcon cannot hear the falconer." But every morning when I see the beauty of the sunset, the fire in the sky reminds me of what we do not do or speak of. Is this the apocalyptic event that Yeats spoke of? Our hubris and Prometheus's gift, without thought of Epimetheus's counsel, have turned us all into arsonists.

Resilience
Angela Menzies

12 x 12 x 1.5 inches

Acrylic on cradled panel

Wolves are known to endure the aftermath of forest fires with quiet resilience. In the face of habitat devastation, the wolf adapts, reroutes, and survives – moving through charred terrain, landscapes reduced to ash, and unfamiliar pathways. Wolves constantly remind us that survival is dependent upon deliberate endurance and the will to remain.

Burning down the House
Kim Money

18 x 18 x 1.5 inches

Acrylic paint, spray paint, crayon and charcoal on cradled panel

I always wondered what the Talking Heads were referring to with their Burning Down the House lyrics, so I looked it up. It is a metaphor for destroying something that has entrapped you—such as your mortgage.

Firestorm

Tony Bounsall

16 x 20 x 1.25 inches

Pigment ink on Hahnemühle Photo Rag

Firestorm is meant to evoke the primal fear and awe fire can inspire. As an elemental force, fire consumes, destroys, and reshapes everything in its path. It is both a cleanser and a harbinger of devastation. My composition focuses on the swirling, dynamic vortices of flame and smoke—impossible to fully capture, yet mesmerizing in their fluidity. Through colour and movement, I convey the violent beauty of wildfire, reminding viewers of fire's indiscriminate and inevitable power.

...AND YET
GLENDA BALKAN

16 x 20 x 0.25 inches

Oil on canvas

The figure in this piece metaphorically suggests the effects of karma. The impassive expression whilst being engulfed by flames exposes our environmental neglect as well as its inevitable outcome. And yet... while the face of this androgynous figure is lit by the glowing heat, it remains whole, beautiful and at peace. And yet...a boundary has been drawn.

Smoke and Flame

Roger Champagne

6 x 7 inches

Ceramics

As a Victoria-based artist I draw inspiration from the West Coast's natural beauty, incorporating its vibrant colours and organic shapes into my pots. My raku works exude a responsive quality, allowing the clay to participate in the end result. Each piece is a manifestation of my imagination, reflecting the endless possibilities that come with working with such a versatile medium. Hopefully my pottery will invite viewers into a space where they may contemplate and reflect on the boundless potential of this medium.

The Campfire

Sandra Curran

30 x 24 x 1.5 inches

Acrylic on canvas

I carry fond memories of fire, from the use of sparklers as a small child at Halloween and New Year's celebrations to the many summer evenings spent around campfires with my ten siblings, sharing stories while roasting marshmallows and hot dogs. I was always captivated by the flickering flames of the campfire, watching the different colours appear as the fire grew and we would shuffle our chairs to avoid the smoke and sparks. We would take turns stomping out the wayward sparks ensuring the nearby underbrush did not catch on fire and enjoying the warmth of the fire after an evening swim in the lake.

Influence

Andrew Bradford

12 x 16 inches

Pigment ink on archival cotton paper (1/25 limited edition)

A single tendril brings life in this charred place. So small, yet soon this landscape will drip with moss.

Campfire Glow
Tony Bounsall

16 x 20 x 2 inches

Mixed media, pigment ink on Hahnemuhle Rag, polyester resinmatte finish

In contrast to my piece, Firestorm, Campfire Glow embraces fire's gentler, more intimate side. This work captures the delicate warmth and flickering light of a campfire on a quiet night beneath the trees. This is a digital pigment print on Hahnemuehle artist rag paper: panel-mounted and finished with a poured polyester resin surface, hand-worked to a matte finish. This process lends a tactile depth and a subtle glow that mimics the soft radiance of firelight piercing twilight.

Friend and Foe

Pierrette Vézina

12 x 12 x 1.5 inches

Acrylic paint and collage on cradled canvas

Fire is like art in that contrast is key. Fire can be a source of comfort or a source of terror. It is bright, but its smokey byproduct is dark. Fire can bring joy or sorrow. Fire is the flip of a coin.

When Prometheus Stole Fire

Phyllis Schwartz and Edward Peck

24 x 18 inches

Analogue collage

Rooted in the myth of When Prometheus Stole Fire, this analogue collage approaches the story through ideas that linger in the charged instant when a forbidden force leaves the realm of the gods and becomes a human inheritance. Fire appears only as colour-shapes—reds, purples, oranges, yellows—that suggest heat, risk, and possibility without resolving into literal flame. These forms hover between gift and transgression, a reminder that some decisions cannot be undone. When Prometheus Stole Fire is a meditation on the threshold and the unsettling paradox of protection and peril.

Epilogue

Kathy Daley

24 x 24 x 1.5 inches

Oil on wood cradled panel

Mother Earth is in trouble. And fire is her messenger. While the global climate crisis has seen millions of hectares of forests razed in the last year alone, it's a source of wonder. We all continue to lead our lives in the face of this ecological doom. I explore this point by putting distracted people in harm's way. The tension plays out in thick impasto oil paint and vivid colours. It's a five alarm fire that begs to be seen, heard and acted upon. Our survival depends on it.

Mystic Meditation
Khione Gardner

36 x 18 inches

Acrylic on canvas

Mystic Meditation

Full faced Moon
Clouds race and dance
Rhythmic waves
Salt breeze caresses
Crackling flames, sparks fly Smoke whispers
Pages curl and burn
A heart begins to heal

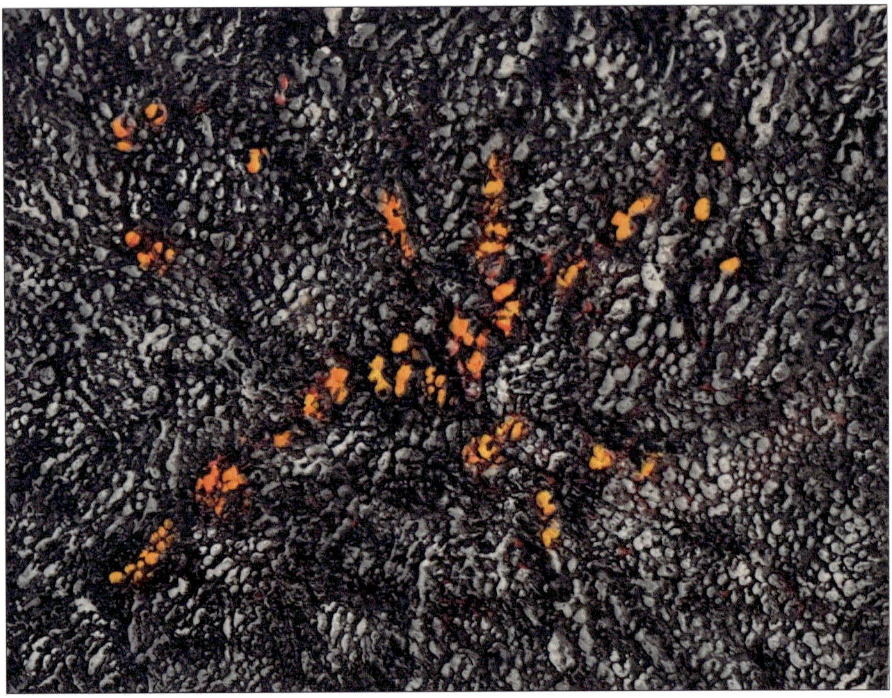

Coals Ready to Reignite (juror's award)

Anita Doornekamp

16 x 20 x 1.5 inches

Encaustic painting on cradle board

This encaustic work simmers with a muted intensity. The carbon-black of the coals and the patches of burnt orange pulse beneath a veil of ash. A single breath might coax the fire to life once more. There is suspense and expectation, along with a charged silence before the flames flicker to life. The textures, layers and colours evoke the potential of the coals. Does the viewer feel a foreboding sense of hidden ferocity? Is it a promise of a controlled, comforting and hypnotic burn? Is the viewer anticipating a sacred rebirth from a time of hardship, a rising from the ashes? This representation of smouldering coals is created by fire; the process is encaustic accretion, the repetitive build-up of layers of wax to create rich, organic, 3-D textures.

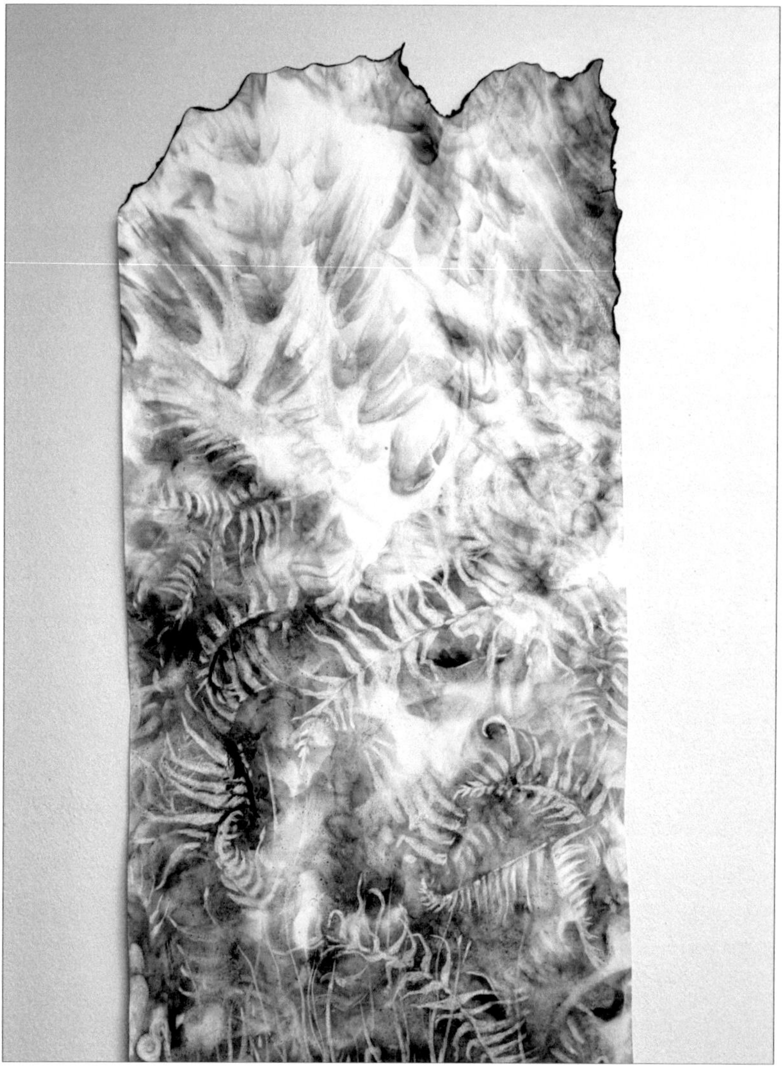

Fuel Layers

Debra Gloeckler

56 x 18 inches

Soot, erasure on paper

My work reflects on climate-driven wildfires intensified by fossil-fuel emissions. Ferns link prehistoric forests, ancient carbon now burned as fuel, to the dry understory that carries fire through forests today. Drawn in soot through erasure, the forms hover between flame and new growth, tracing ecological change, regeneration, and resilience in fire-adapted ecosystems.

Witnessing the Approach of the Fire III

Jean Slick

24 x 36 x 1 inches

Acrylic on canvas

At one point during the Lost Creek wildfire in the Crowsnest Pass, Alberta, smoke had spread across the southern horizon near my weekend home, blocking daylight and dropping ash across everything in sight. At times flames were visible; at others they glowed behind a dark, yet sometimes translucent wall of smoke. This painting holds that experience: the way smoke blocked the light from most of the sky while both revealing and concealing the fire. Witnessing its growth brought acute awareness of fire's immense power and the community's vulnerability. Throughout history, painting has been one way to witness disasters beyond the moment of their occurrence. My work draws on my photographs taken during the fire with the act of witnessing extending from in-person, to lens, to brush, to gallery, where others become secondhand witnesses. In my research on wildfire and social media, a common theme was the sharing of citizen-generated images of the approach of a fire.

Fire's Parting Kiss

Phyllis Schwartz

16 x 20 inches

Pigment ink on fibre

Fire's Parting Kiss depicts the quiet passage from day to dusk, the horizon a gentle threshold holding the last warmth of the setting sun. Calm water and fading light create an atmosphere of stillness and reflection. An anchored sailboat serves as a silent witness to the sky's final ember, its mast a narrow line of stability within the sea's slow movement. The transformation is subtle: a sibilant flame gradually dissolves into muted grey, tracing the day's departure. Light and shadow intermingle across water and sky, adding depth and a restrained tension that suggests both serenity and slow transition. A faint, distant ship introduces a sense of distance, memory, and unseen journey, extending the landscape beyond the visible and holding the final warmth of the sun's fire within the cool approach of dusk.

Primordial Cave
Margo Farr

11 x 13 x 10 inches

Plaster, beeswax, dyed wet-felted fleece, sticks, stones, bones

My inspiration when creating this piece was to represent elements of nature from the very beginning of time - the forces of fire, stone, wood, and animals in earliest cave environments. The materials were chosen to embody each of these elements, which were essential for survival. Our human relationship to these forces has transformed through millennia.

Damaged Not Destroyed

Valerie Dickson

8 x 10 x 0.25 inches

Oil on canvas panel

This painting was inspired by the capricious nature of fire, and the ability of fire-ravaged landscapes to regenerate.

The Fire
Kathy Daley

36 x 36 x 1.5 inches

Oil on wood cradled panel

Mother Earth is in trouble. And fire is her messenger. While the global climate crisis has seen millions of hectares of forests razed in the last year alone, it's a source of wonder we all continue to lead our lives in the face of this ecological doom. I explore this point by putting distracted people in harm's way. The tension plays out in thick impasto oil paint and vivid colours. It's a five alarm fire that begs to be seen, heard and acted upon. Our survival depends on it.

Paddy's Pond

Judy Taylor

11 x 14 x 1.5 inches

Cold wax and oil on canvas

The summer of 2025 was my first exposure to a summer of fire. Newfoundland is a rock in the Atlantic Ocean, surrounded by salt water, and the land is dotted with wet marshes, gullies, ponds and a few very large fresh-water lakes. By August, snap, crackle and pop were underfoot ready to ignite. Shades of peach, orange, pink and red graced the skies creating memorable sunsets and sunrises. Yes, the fires were destructive. There was a loss of property, evacuations, and rancid smoke in toxic colours. A photo of the Paddy's Pond fire taken from the passing car. Conception Bay fires, smoke visible for 50 miles, destroyed structures, homes, business and cabins. I saw a challenge—a painting challenge—the unseen destruction through the imagery of smoke and flames.

Fawn Lilies Return
Joanne Thomson

28 x 42 x 2 inches

Recycled wood, acrylic, casein, fire, wax

Fawn Lilies have deep bulbs. They were safe under the forest floor when the fire swept through the forests near Sooke Potholes. They came back the first year. Fire may think it is powerful. Yet, quietly waiting in the damp soil underneath are those who will survive.

FIRST THEY COME FOR THE BOOKS
DIANA SMITH

30 x 24 x 0.75 inches

Charred paper, collage papers, acrylic paint, graphite

I stumble across burnt pages of a book. Later I read an article about the Canadian government plans to axe prison librarians, defying Rule 64 of the Nelson Mandela rules regarding minimum treatment required for prisoners. Knowledge, and concepts of truth and justice are always under threat. Here fragments symbolizing vulnerability offer a warning.

Witnessing the Approach of the Fire I

Jean Slick

28 x 22 x 1 inches

Acrylic on canvas

Both personal and professional connections to wildfire inform my paintings. During the Lost Creek fire in the Crowsnest Pass, Alberta, the first sign of danger wasn't flame - it was sky. Growing pyrocumulus clouds, generated by the intense heat of a fire still hidden behind the mountains near our weekend home signalled the fire's presence before it could be seen. Watching those clouds build and shift, I found myself drawn to the beauty of the colourful, dramatic, and rapidly changing cloud formations, which created a scene that was both ominous and mesmerizing. In my research on wildfire and social media, I found that witness images fall into three categories: watching the approach, fleeing, and watching one's house burn. This painting belongs to the first category - the moment where awe and perception of risk intersect, and where the fire revealed its powerful and dangerous presence in the landscape.

Wild Fire IV

Bill Horne

6 x 9.5 x 0.1 inches

Silkscreen with gold ink (1 of 14)

Wildfires directly impacted Claire Kujundzic and me during several summers when we were living in the North Cariboo. For weeks at a time, we kept a kit of possessions ready to take with us on short notice, while ash and cinders fell from darkened skies. We kept our car's gas tank full, and monitored fire maps. These were the conditions that literally sparked my *Wild Fire* prints. Partially mixed ink colours print in serendipitous streaks due to the nature of the screen printing process, so the skies behind the boreal forest silhouette can vary throughout an edition.

Scorched Symphony

Liz Hughes

14 x 11 inches

Mixed media

This piece is my way of confronting the urgent reality of our changing world. Scorched Symphony emerges from reflections on the wildfires that have been intensified by climate change. The limited palette of reds and yellows depicts the relentless heat and energy of these blazes as they consume the landscape, while the sweeping blackened curve symbolizes the scars left behind - dark reminders of our role in the situation. Through multiple textured layers, I aimed to convey the fierce movement of flames and the tension between destruction and resilience.

Fire+Ash+Clay

Irene Ives

14 x 10 x 8 inches

Cone 10, wood fired stoneware

Humans have harnessed the power of fire to transform clay into ceramics for over 35,000 years. Many potters specialize in atmospheric firing methods using a reduction atmosphere. Combustible fuels use the environment's oxygen causing chemical reactions with the clay and glaze. They are the true alchemists of our art form, taking a lifetime to master these techniques. On Denman Island I participated in a 4-day firing of Gordon Hutchens' *anagama* wood kiln stoking every 15 minutes day and night with a team working in shifts to keep both the kiln and the crew fed. As the kiln temperature creeps higher, the convectional atmosphere swirls ash around and across the clay where it lands and forms a glazed surface while the dancing currents and fluctuating atmosphere leaves unpredictable yet unmistakable marks on the ceramic forms. These woodfired pieces capture the fire-created surfaces that, as candle holders, provide a lasting reminder of the integral role fire has had in their creation.

Wild Fire IV
Bill Horne

6 x 9.5 x 0.1 inches

Silkscreen with gold ink. (4 of 14)

Wildfires directly impacted Claire Kujundzic and me during several summers when we were living in the North Cariboo. For weeks at a time, we kept a kit of possessions ready to take with us on short notice, while ash and cinders fell from darkened skies. We kept our car's gas tank full, and monitored fire maps. These were the conditions that literally sparked my Wild Fire prints. Partially mixed ink colours print in serendipitous streaks due to the nature of the screen printing process, so the skies behind the boreal forest silhouette can vary throughout an edition.

Looking Up

Cathy Hammond

14 x 11 x 1.5 inches

Acrylic on canvas

This painting was created while I was in high emotion worrying about my two daughters and young grandson who were living in Kelowna when Bear Mountain and Knox Mountain caught fire. My youngest, who was at home with her son, started to take a video of West Kelowna and how quickly the fire was heading towards downtown where they lived in an apartment. By the time my eldest daughter got home from work the fire was across from them with only the lake between them. The fire did cross over to Knox Mountain later that evening, and that is when they started to pack their bags in case of an evacuation. I did not sleep too well that night once seeing the video my daughter sent me! The following day, Looking Up came to life. I was grateful that they were safe, but also very sad for all those who lost their homes as well as the animals. I am also truly grateful for all the firefighters who continue to keep our communities safe, while putting their lives in danger for us all.

Contracts forged with fire

Mary Wulff

12 x 15 x 1.5 inches

Acrylic and canvas

We searched the market at Bahir Dar for a metalsmith who could put a clasp on a rare coin. I loved the glow from the flame on their faces. The challenge was to add the clasp without ruining the coin. Coins provide an enduring historical record of cultural values and beliefs.

Embers

Kimberley Zutz

20 x 16 x 1.5 inches

Acrylic and posca markers

When faced with the destruction of fire it seems to be so complete, so devastating, so unstoppable. Forest fires are an annual happening in our world as climate change shows us all how fragile the planet has become; but there are times when there is beauty in the force of a fire.

From the Ashes

Lynn Gordon-Findlay

17 x 23 inches

Mixed media

Fifty years after my family home burned to the ground, “From the Ashes” was created as a cathartic part of finally processing that trauma. The composition is designed in a series of layers to be created then burned away, as if re-enacting past destruction, then overlaid to construct something whole, though evidence of scars remains. The house plan was hand drafted from memory in pencil, ink, and charcoal. The view of the house and surrounding property was drawn in raw charcoal from the only remaining photograph. Three objects salvaged from the fire were rendered in a variety of media on black illustration board as if emerging from blackened ash. Every stroke of pencil, scrape of charcoal, smear of clay brought back long forgotten memories and associated joys. Every charred piece of paper that fell away under flame elicited long buried emotions. Every layer became part of a regenerative process that transformed a devastating loss into a healing transformation.

The Fire at the Heart of AI

Mary Wulff

24 x 30 x 1.5 inches

Acrylics and inks on canvas

I treat google like a pilgrimage to the Oracle of Delphi with its eternal flame. Oracle knows my finances, my food, exercise and sleep routines, my philosophy, my entertainment, my politics. I consult it on everything from recipes for cooking to the likelihood of nuclear war. In this painting I imagined a doctor using AI to do an analysis of me as a patient and the AI having access to all my personal data. Based on the results the doctor is instructed to detain me, that I am not 'sane'. The doctor disagrees with the AI and says no... he can't detain me based on my search history. He reaches to pull the plug on the AI but AI reminds him that his heart pacemaker is AI controlled. He is helpless to do anything other than what AI instructs him to do. Artificial intelligence may yet enslave us with its demands to feed the fires of the eternal flame that maintains it. This painting is a cautionary tale that contains references to E. Delacroix's painting of Lycurgus consulting Pythia.

Fire+Clay+Salt

Irene Ives

12 x 18 x 8 inches

Cone 10, salt-fired stoneware

Humans have harnessed the power of fire to transform clay into ceramics for over 35,000 years. Many contemporary studio potters specialize in atmospheric firing methods using a reduction atmosphere. Combustible fuels use the environment's oxygen causing chemical interactions with the clay and glaze. They are the true alchemists of our artform as it takes a lifetime to master these techniques. Learning to harness the power of the fire to coax desired, yet always unpredictable, outcomes is part of the allure. Whenever possible, I take the opportunity to learn from these masters. Salt-firing has deep roots in ceramic tradition; throwing salt into the kiln at exactly the right times so that it vaporizes and lands on the ceramic surfaces creating the trademark glossy, translucent orange-peel surface. These salt-fired forms capture the fire-created surfaces that, as candle holders, provide a lasting reminder of the integral role fire has had in their creation.

Index of Artists

Endnotes

Metchosin ArtPod gallery is located in Metchosin, British Columbia. We endeavour to support local visual artists by offering opportunities to learn, create, show, and sell original work, and to promote the health of our community by providing a venue for the public to engage with the arts.

Using a blind jury process, independent juror Samantha Dicke selected works that interpreted the theme of the exhibition. In addition, the juror exhibited their own works.

The exhibition was curated by Khione Gardner, Kim Money and Diana Smith, and ran from March 20 to May 3, 2026. This exhibition book features all works juried in the exhibition and can be purchased from Amazon's online bookstore. Copies of the book are held in Library and Archives Canada.

The installation designer was Khione Gardner and a team that included Irene Ives, Anita Doornekamp, Angela Menzies Anna Mar and Kim Money.